Patterns and Inferential Networks

Thomas Gil

T0119054

Philosophische Hefte

Band 11

Herausgegeben von

Prof. Dr. Axel Gelfert
Prof. Dr. Thomas Gil

Patterns and Inferential Networks

Thomas Gil

Logos Verlag Berlin

λογος

Philosophische Hefte

Herausgegeben von

Prof. Dr. Axel Gelfert
Prof. Dr. Thomas Gil
Institut für Philosophie, Literatur-, Wissenschafts- und
Technikgeschichte
Technische Universität Berlin

Bibliografische Information der Deutschen
Nationalbibliothek

Die Deutsche Nationalbibliothek verzeichnet diese
Publikation in der Deutschen Nationalbibliografie;
detaillierte bibliografische Daten sind im Internet über
http://dnb.d-nb.de abrufbar.

ISBN 978-3-8325-5057-8
ISSN 2567-1758

Logos Verlag Berlin GmbH
Comeniushof, Gubener Str. 47,
10243 Berlin

Tel.:   +49 (0)30 / 42 85 10 90
Fax: +49 (0)30 / 42 85 10 92

http://www.logos-verlag.de

Contents

Preface

If God or very tiny animals asked ontological questions the answers they would find would differ significantly from the answers human beings have given to the question concerning what there is. Ontology seems to be a relative business ("scale relative"), depending on who asks questions, on what level the questions have been asked, and what are the theoretical interests motivating the questions.

Human beings seem to be prone to talk and think of objects. They persist in breaking reality down into a multiplicity of identifiable and discriminable objects. The groups and societies they belong to coach them alike in a pattern of verbal response to externally observable inputs. Socialized in such a way, they converge identifying matter and individuating it. In other words, they learn to master the conceptual scheme of individuation and enduring physical objects getting along thus with other people and coping with external reality. Most of the time, practical success confirms their ontological decisions.

Even if science seems to be more interested in relations, correlations, patterns and structures, these scientific entities would not come about without single objects and their (characteristic and dispositional) properties. What actually happens is due to what there is and its dispositional qualities. And the patterns and correlations that interest science seem to derive from specific material things behaving in typical ways in specific situations and constellations.

Part one is dedicated to individual objects, their relations, and the patterns that come about when they relate and correlate. Part two is about knowing what there is, that is, experiencing something and forming hypotheses about the experienced and its ways of functioning. Our hypothetical thinking about reality is not atomistic. We accomplish it creating inferential networks that make possible transitions and inferential connections. And when we introduce the formal notational language of mathematics to express and promote it, our structural knowledge of what there is acquires a degree of exactness and precision not accessible without such abstract formality.

Even if our tendency to reify leads us in a straight way into an ontology of individual material things and individual happenings and events, we should not forget that things can only actually happen due to the qualities and properties of individual objects related to other objects in specific situations. When mathematically grasped in formal structures, situated individual things and events become positions in real patterns that make up reality. Such objective patterns are the referential domain of mathematical formulae, equations and structures. Things and events, this is the conclusion of the first part, exist pattern-like. Pattern-like too is the way we reason going inferentially from informational bases to hypotheses, confirmations, and conclusions.

Wilfrid Sellars distinguished the "scientific image" from the "manifest image" of "man-in-the-world". Taking up such a wise distinction, we could say that, according to the "manifest image", individual material things and events are the constituents of reality while, according to the "scientific image", reality is a set of objective patterns in which existing things and events are positions relevantly related to other positions.

1 Patterns

1.1 Objects and Relations

The question concerning what there is in the world is the ontological question, that is, the question about the world, the question about reality. An easy way of answering it would be to say that there are things and that that is all there is. But how are we to understand those things? Are they substantial individuals or particulars? Or, simply, things that happen and are to be further characterized using the conceptual and theoretical devices we estimate to be appropriate? We refer to the things there are in the world when we speak and we say something about them. Speaking and describing them, we identify them as individuals or particulars. When we say "the plant grows", "the cake is on the table" or "Peter smokes" we identify single things, say something about them, describing them and ascribing them characteristic features. We introduce them as being somehow existent, this identifying introduction being the first part of making a statement about them. Put differently, through definitely identifying expressions we introduce particulars or individuals in a sort of "naming game" without necessarily possessing the conceptual resources for properly identifying the corresponding particulars.

We presuppose simply some sort of existence for the things we refer to in our "naming game".

But are those things we refer to when we identify and name them substantial entities or substances, as Aristotle would suggest? In a severe anti-Platonic mood, Aristotle characterized the things there are as autonomous beings of which we can say something essential or accidental. What is "said of" them is essential to them, according to Aristotle, while what is said to be "in" them is accidental to them. One thing is fundamental in Aristotle's picture. The things about which we say something cannot be said about something else. They cannot be "said-of" nor "said-in", they are "primary substances" ("prôtê ousía"). For Aristotle, such substantial things exist in a basic, non-derived, and independent way.

We can agree with Aristotle and say that where there are qualities we predicate of things, there are things qualified, and that the things thus qualified may sustain qualitative change while going on being or remaining (ontologically) one and the same. And we can further agree with Aristotle that when the world presents us with the fact of a white Socrates, the right thing to say is "Socrates is white" and not "Whiteness is Socrates" or "Whiteness Socratizes". To this extent, we can go Aristotle's way.

However, an ontology consisting of (essentially and accidentally qualified) substantial things ignores relevant features of reality. In Plato's, David Lewis' and Theodore Sider's terminology, an ontology of substantial things does not properly "carve at reality's joints", matching adequately reality's structure and its fundamental features. In addition to that, an ontology consisting of substantial things creates several unnecessary and unjustifiable distortions, not easily to be eliminated, even if it may be, at first sight, intuitively plausible. Making too many problematic unnecessary assumptions, it violates the principle of parsimony that prescribes to assume only what is strictly necessary.

Traditionally, qualities and properties have been attributed to existing things when these had to be characterized or simply classified. The task of scientific research would be, according to such a view, to find out what qualities and properties should be predicated or said of the things in question. How those qualities and properties had to be predicated was important too. Then not all qualities or properties are predicated in the same way. Some seem to be essential for the identities of the things studied, while others are merely accidental and not necessary. If we say of a certain cat, to give a very simple example, that it is an animal, the quality or property of "being an animal" is an essential quality or property that can be attributed to all single cats. Other qualities and properties may be of this individual cat, for instance, the property of "being black". But this property is not a necessary property that can be attributed to all cats. There are many cats which are not black and do not have that property. That means, not all cats instantiate or exemplify the property of "being black". Therefore we call such a quality or property an "accidental" property.

There are some properties that can be reduced to other properties. Consider a situation which arises when a scientist asserts that temperature is mean molecular kinetic energy. What is being asserted is that the property of having a particular temperature is really the same property as the property of having a certain molecular kinetic energy. Such a case is a classical example of a reduction of one physical property (or magnitude) to another physical property.

Having a property often amounts to having a certain power or disposition. In some cases the only informative things we can say about a property are what powers, capacities or dispositions it confers on its instances. Properties as dispositions make possible that things happen. They have causal efficacy that becomes manifest under appropriate conditions. Dispositions are important due to two main reasons. They let world structure become manifest when they actually cause something, as they show in a very concrete way that things happen always relationally. Disposition manifestation is namely always something involving properties and situations or stimulus conditions. "Fragility", "solubility, "malleability" (the typical examples given, when authors have to say what it is to be a disposition) are dispositions, all of them structurally relational. Things are fragile, soluble or malleable only in certain situations, contexts and conditions. And, this would be the second reason why dispositions are important, disposition ascriptions or attributions cannot be analysed in an exclusively extensional language. Intensions and counterfactual or subjunctive conditionals are always involved whenever we attribute dispositional predicates, as saying, for instance, that "x is soluble at t" means that "x would dissolve if put into water at t". Dispositions as powers, abilities, potencies, capabilities, tendencies, potentialities, proclivities or capacities, always linguistically (that is, intensionally and modally) expressed, show paradigmatically the relational structure of reality.

"Relations" can be seen as a specific kind of properties (two-, three-, or four-place properties). Being the father of somebody, for instance, could be then formalized as a dyadic property that characterizes someone. However, the logical formula we use to represent symbolically relations with two "relata" R (a, b) could be interpreted more fundamentally, letting relations appear as a main "structure" that allows things to be what they are. Being related would not be to possess an additional quality or property, the property of being related. Being related would be the only way for something to be really.

Dispositional properties manifest real world structure. That is why they can be referred to in scientific explanations. Like dispositions, certain properties, relations and happening things manifest also world structure. They are more than conceptual and theoretical tools we have fabricated to speak about reality. They exist really. They are "ontic structure", and not merely the products of functional, explanatory language. However, we need language, that is, concepts, propositions and theories in order to individuate and describe them, i.e. in order to say what kind of ontic structure they are. Alternative ways and manners of individuating or describing real world structure will always exist. This does not mean that our languages and idioms create the things they speak about. Our linguistic individuation and description strategies may vary and change. But they cannot be completely arbitrary as they are structurally constrained by "reality's joints".

What is important here is to stress that so-called event-sentences reveal real world structure, and that they are not arbitrary creations of our grammars. Event-sentences say what kinds of things happen due to different properties of individual things acting and reacting in specific situations. The world is indeed made of things, properties, happening things and occurrences that are relationally structured and linguistically identified and described.

1.2 Patterns

Must all the things we are acquainted with in daily life go in scientific ontology, as James Ladyman's and Don Ross' expression "Every Thing Must Go" prescribes? Not really. Ontology is scale relative. Scale relativity means that terms of description and principles of individuation we use to track the world vary with the scale at which the world is grasped. In everyday life, there are cars, tables, chairs, houses, people, cats and dogs. At the quantum scale there are no cats, no dogs, no persons, no tables, and no houses. At scales appropriate for astrophysics there are no mountains, no hills, and no valleys.

Sciences are interested in patterns, in real patterns that help us explain what there is and what happens. Adopting the perspective of our best scientific theories (quantum mechanics and relativity theory), we may however identify "real patterns of high indexical redundancy" and "unusually strongly cohesive real patterns" (so effective at resisting entropy that we can transport them to new environments and thus relocate them) (Ladyman, Ross, 294).

Structural scientific language allows us to refer more adequately to subsisting individual objects, to "natural kinds" (such as water, gold, and silver), and to occurring things and events. These objects, "natural kinds", and events become "locators" of real patterns from which structural science may extract information via measurements and other tracking devices.

Patterns really exist. They are the things that count for the special sciences. But in everyday life the totality of different relevant so-called "locators of real patterns" is all that matters.

Modern science forces us to abandon traditional ontology, the ontology of particularism which assumes that: 1. there are individuals in space-time whose existence is independent of each other; 2. each such individual has properties that are intrinsic to it; 3. relations between individuals supervene on the intrinsic properties of their relata; 4. the identity and individuality of material objects can be accounted for mainly in qualitative terms.

Both quantum mechanics and relativity theory teach us that the nature of space, time, and matter challenges radically traditional ontology that describes the world as composed of subsistent particular material objects. However, it is such a primitive ontological "particularism" that individual organisms like human beings presuppose in their daily efforts to live and survive. For structural scientists the only existing entities are "real patterns" behaving like things or like events and processes. In their daily book-keeping practices, however, human beings reasonably take for granted the real existence of things, events and processes.

2 Inferential Networks

2.1 Inferences

Inferring, when reasoning, we move from premises (certain initial statements) to logical consequences or conclusions. Traditionally, inferences are divided into deductive, inductive and abductive inferences. Deductive inferences derive logical conclusions from premises known or assumed to be true. Inductive inferences move from particular premises to universal conclusions. Abductions are inferences to the best explanation. Statistical inferences, ever more important in contemporary science, use mathematics to draw conclusions in the presence of uncertainty. With or without mathematics, the most interesting inferences are the ones pertaining to inductive reasoning through which we get from multiple observations to conclusions.

The validity of deductive inferences (or arguments) is determined by their logical form, not by the content of the statements involved. That means that their validity depends solely on the relation between the premises and the conclusion so that necessarily the conclusion must be true if the premises are true. Validity is thus a property of inferences or arguments, which are groups of statements, not of individual statements. In other words, form or structure is what counts. Validity is determined by the form, not by what the premises and conclusion refer to.

Unlike deductive arguments, inductive inferences and arguments provide conclusions whose content exceeds (or goes beyond) that of their premises. To do this, they have to sacrifice the necessity of deductive arguments. Therefore, if the premises of a correct inductive argument are true, the best we can say is that the conclusion is probably true. In other words, the premises of a correct inductive argument may render the conclusions extremely probable, moderately probable, or probable to some extent. Additional evidence may have crucial importance for inductive inferences.

Abductive inferences are a form of hypothetical reasoning which starts with certain observations and seeks to find the simplest and most likely explanation for them. The reasoning process, unlike deductive reasoning, yields only plausible conclusions. Abductive conclusions are qualified as having a remnant of doubt or uncertainty, something that is expressed by terms such as "best available" or "most likely".

The power of science rests in its ability to establish, on the basis of observational evidence, far-reaching hypotheses about reality. Some hypotheses are universal generalizations, other hypotheses are statistical generalizations. When scientists speak about the reasonableness or plausibility of hypotheses, they are always assessing probabilities, in rough estimates or using precise mathematical apparatus. But it should be re-emphasized that the whole business of confirmation of hypotheses is always inductive. This means that scientific hypotheses are never completely verified or absolutely true.

2.2 Inferential Networks

Mathematical thinking is not contemplation of ideal objects or forms existing outside the empirical world. Mathematical thinking is structural thinking using formal tools and devices.

Mathematical knowledge has its roots in pattern recognition and pattern representation. However, it is important to stress that knowledge of a pattern is quite different from pattern recognition. Animals in their daily struggle for survival are able to recognize (simple) patterns, but they certainly cannot represent and describe such patterns. The primary subject matter in mathematics is the structural arrangement of individual positions. Michael D. Resnik prefers the expression "... that the objects of a mathematical theory are the positions in the structures it describes" (Resnik, 218). Such a formula indicates Resnik's epistemic reading of structuralism according to which there is not structuralism all the way down: "Realism about mathematical objects does not commit one to realism about structures, even when one maintains that mathematics studies patterns or structures and that mathematical objects are positions in patterns" (Resnik, 261).

Mathematics is, indeed, knowledge of patterns and knowledge of positions in patterns. Take as an example Galileo Galilei. Instead of trying to find an explanation of what causes an object to fall to the ground when released from the top of a tower, Galileo tried to find out how the position of the falling object varies with the time since it was dropped. And he discovered that the distance travelled by a ball at any instant varies with the square of the time of fall. Using modern algebraic terminology, he discovered the relationship $d = kt^2$ that connects the distance of fall d with the time of fall t, where k is a constant.

Galileo identified certain features of the world that can be measured, and looked for meaningful relationships between those features. Galileo concentrated in features like time, length, area, volume, weight, speed, acceleration, inertia, force, moment, and temperature. He ignored colour, texture, smell, and taste. Or consider Isaac Newton's well-known law of force: The total force on a body is the product of its mass and its acceleration ($F = mXa$).

This law provides an exact relationship between three highly abstract phenomena: force, mass, and acceleration.

David Hilbert's "Foundations of Geometry" could be a paradigmatic example of what it is to think structurally in mathematics. For Hilbert, "points", "lines" and "planes" are not absolute entities, each of them definable independently from the others. They are considered to be elements of three different "systems". The first system is composed of "points" (A, B, C ...). The second system is composed of "straight lines" (a, b, c ...). The third system is composed of "planes" (α, β, γ ...). And then, "points", "lines" and "planes" are thought to have certain mutual relations, indicated by means of such words as "are situated", "between", "parallel", "congruent", "continuous", etc. Five groups of axioms (axioms of connection, axioms of order, axioms of parallels, axioms of congruence, and axioms of continuity) determine stipulatively in a complete and exact description what those mutual relations are.

Like mathematical thinking, our inferential activity is also to be conceived of structurally. Our inferences are always related and connected to other inferences in "inferential networks". Inferring is not an atomistic enterprise. It takes place "holistically" in networks in which inferences cohere with other inferences, supporting other inferences, and being supported by them.

Bibliography

Allen, S. R., A Critical Introduction to Properties (London: Bloomsbury 2016).

Bealer, G., Quality and Concept (Oxford: Clarendon Press, 1982).

Bennett, J., Events and Their Names (Oxford: Clarendon Press, 1988).

Chalmers, D. J., Manley, D., Wasserman, R. (Eds.), Metametaphysics. New Essays on the Foundations of Ontology (Oxford: Clarendon Press, 2013).

Davidson, D., Essays on Actions and Events (New York: Clarendon Press, 1989).

Drake, S., Galileo. A Very Short Introduction (Oxford: Oxford University Press, 2001).

Heil, J., The Universe as We Find It (Oxford: Clarendon Press, 2015).

Hempel, C., Aspects of Scientific Explanation and other Essays in the Philosophy of Science (New York: The Free Press, 1970).

Hilbert, D., The Foundations of Geometry (Merchant Books, 2007).

Hirsch, E., Dividing Reality (Oxford: Oxford University Press, 1997).

Iliffe, R., Newton. A Very Short Introduction (Oxford: Oxford University Press, 2007).

Ladyman. J., Ross, D., Spurrett, D., Collier, J., Every Thing Must Go. Metaphysics Naturalized (Oxford: Oxford University Press, 2010).

Lewis, D., Papers in Metaphysics and Epistemology (Cambridge: Cambridge University Press, 1999).

Lipton, P., Inference to the best Explanation (London: Routledge, 1991).

Mellor, D. H., Oliver, A. (Eds.), Properties (Oxford: Oxford University Press, 1997).

Mumford, S., Dispositions (Oxford: Oxford University Press, 1998).

Psillos, S., Scientific Realism. How science tracks truth (London: Routledge, 1999).

Quine, W. V. O., From a Logical Point of View. Nine Logico-Philosophical Essays (Cambridge, Massachusetts: Harvard University Press, 2nd edition, 1980).

Resnik, M. D., Mathematics as a Science of Patterns (Oxford: Clarendon Press, 1999).

Salmon, W. C., Logic (Englewood Cliffs, N.J.: Prentice Hall, Inc., 1963).

Salmon W. C., Causality and Explanation (New York: Oxford University Press, 1998).

Sellars, W., Science, Perception and Reality (Atascadero, California: Ridgeview Publishing Company, 1991).

Shields, C., Aristotle (London: Routledge, 2nd edition, 2014).

Sider, T., Four-Dimensionalism. An Ontology of Persistence and Time (Oxford: Clarendon Press, 2010).

Sider, T., Writing the Book of the World (Oxford: Clarendon Press, 2011).

Strawson, P. F., Individuals. An Essay in Descriptive Metaphysics (London: Routledge, 1964).

Tahko, T. E., An Introduction to Metametaphysics (Cambridge: Cambridge University Press, 2015).

van Fraassen, B. C., The Scientific Image (Oxford: Clarendon Press, 1980).

van Inwagen, P., Material Beings (Ithaca, NY: Cornell University Press, 1990).

Bandaufstellung

Alle erschienenen Bücher können online beim Logos Verlag Berlin bestellt werden (https://www.logos-verlag.de).